50 THINGS TO KNOW ABOUT BECOMING AN ENGINEER

Arvindhra Rao Krishnamurthy

Cover designed by: Ivana Stamenkovic
Cover Image: https://pixabay.com/photos/technical-drawing-calipers-workshop-3324368/

CZYK Publishing Since 2011.

50 Things to Know

Lock Haven, PA
All rights reserved.
ISBN: 9798642461419

50 THINGS TO KNOW

BOOK SERIES
REVIEWS FROM READERS

I recently downloaded a couple of books from this series to read over the weekend thinking I would read just one or two. However, I so loved the books that I read all the six books I had downloaded in one go and ended up downloading a few more today. Written by different authors, the books offer practical advice on how you can perform or achieve certain goals in life, which in this case is how to have a better life.

The information is simple to digest and learn from, and is incredibly useful. There are also resources listed at the end of the book that you can use to get more information.

50 Things To Know To Have A Better Life: Self-Improvement Made Easy!

Author Dannii Cohen

_____

This book is very helpful and provides simple tips on how to improve your everyday life. I found it to be useful in improving my overall attitude.

50 Things to Know For Your Mindfulness & Meditation Journey
Author Nina Edmondso

_____

Quick read with 50 short and easy tips for what to think about before starting to homeschool.

50 Things to Know About Getting Started with Homeschool by Author Amanda Walton

I really enjoyed the voice of the narrator, she speaks in a soothing tone. The book is a really great reminder of things we might have known we could do during stressful times, but forgot over the years.

Author HarmonyHawaii

---

50 Things to Know to Manage Your Stress: Relieve The Pressure and Return The Joy To Your Life

Author Diane Whitbeck

There is so much waste in our society today. Everyone should be forced to read this book. I know I am passing it on to my family.

50 Things to Know to Downsize Your Life: How To Downsize, Organize, And Get Back to Basics

Author Lisa Rusczyk Ed. D.

Great book to get you motivated and understand why you may be losing motivation. Great for that person who wants to start getting healthy, or just for you when you need motivation while having an established workout routine.

50 Things To Know To Stick With A Workout: Motivational Tips To Start The New You Today

Author Sarah Hughes

50 THINGS TO KNOW ABOUT BECOMING AN ENGINEER

BOOK DESCRIPTION

Do you want to know what personality traits should an engineer possess? Would you like to know how the life of a working engineer and engineering student looks like? Do you want the tips on how to tackle the obstacles that is faced by an engineer and engineering student in their life? If you answered yes to any of these questions then this book is for you...

50 Things To Know About Becoming An Engineer by author Arvindhra Rao Krishnamurthy offers an approach to how to deal with engineering life as an engineering student as well as the working world of an engineer. Most books on things to know about becoming an engineer tell you on the personality traits that should be possessed by an engineer and perhaps a little bit of insight into the world of engineering. Although there's nothing wrong with that, they only mention things to know and tips from a general point of view and not the first-hand experience from the eyes of an engineering student or engineer which makes it not so relatable. This book will provide you tips added with the first-hand experience from an engineering student who lived that life to make it relatable to the person reading it.

Based on knowledge from the world's leading experts, the books full of tips are more interesting and exciting to be read when the personal adventures and experiences of the author are added in together in the book. In these pages, you'll discover the things to look out for throughout the engineering life as an engineering student as well as a working engineer. This book will help you get through engineering life in a much easier way since a lot of tips have been added on how to handle the obstacles and hurdles that are thrown continuously in the life of an engineer.

By the time you finish this book, you will know how to tackle the engineering life like a legend. So grab YOUR copy today. You'll be glad you did.

TABLE OF CONTENTS

19. Continuous Learning

20. Curiosity

21. Honesty is the Best Policy

22. Be Humble

23. Consistency

24. Intelligence Always Have To Come Out On Top

25. Socializing

26. Building Connections

27. Engineering Degree Is Indeed Hard

28. Pace Is Way Too Fast

29. Reduction In Guarantee For Jobs

30. Setting Low Expectations On Salary

31. GPA Does Not Matter As Much As You Think

32. Constantly Being Asked To Fix Things

33. Being Asked Ridiculous Questions

34. Feeling Stupid Most Of The Time

35. Final Year Project Is A Complete Mayhem

36. Unsteady Relationships

37. Sleep Cycle Is Going To Be Haywire

38. Capacity To Venture Out Of Engineering

39. Youtube Tutorial Videos Are Complete Life-Savers

40. Lab Times Are Very Important

41. Eating Done Right

42. Do Not Be Afraid To Talk About Your Mental Health

DEDICATION

This book is dedicated to all the people who are working as engineers out there and changing the world to a much better place and to those engineering students who took a bold decision to take up this course. All of you are true legends and keep rocking the world as we always do. One step at a time and we will rule the world in no time.

This book is also dedicated to my parents who supported me in my journey as an engineer as well as an author. I would not be where I am today without my help. I would also dedicate this book to my friends, Jennisha, Dharshini, Brahma and Rakchana who did engineering together with me and helped out in giving out helpful feedback to improvise this book.

ABOUT THE AUTHOR

Engineers are the individuals who can shape the world and bring in new innovation to help in improving the quality of lives.' One such individual involved in this noble engineering course is this author, Arvindhra Rao Krishnamurthy.

He graduated with a Bachelor in Engineering (Honors.) from Universiti Sains Malaysia, one of the top-tier universities in Malaysia. He specialized in Mineral Resources Engineering which deals with the mining and quarry industry. He is well-versed with the engineering life of a student and a working engineer since he was involved in both lives' before. He has a lot of friends who are working as engineers so he wrote this book based on the personal experiences of himself and the people he knows about.

He is currently pursuing his passion for writing and getting his fitness back on track after suffering from a back injury two years ago. You can find him on a few social media platforms. In Facebook, he uses his own name, Arvindhra Rao Krishnamurthy and he uses the username arvindrocker96 on his Instagram account. He gives out movie reviews on his Instagram page since he is an avid movie lover and spends most of his free time with them.

"I don't spend my time pontificating about high-concept things; I spend my time solving engineering and manufacturing problems."

— Elon Musk

As Elon Musk mentioned, that's exactly what engineers do. Engineers are not superhumans. They make mistakes as well in their assumptions and decisions but they rectify it as fast as possible and keep trying till they get the right formula to make things work. They put their ideas into action rather than just pondering about them. That's how legends are born.

For a pessimist, he views the glass as half empty while an optimist views the glass as half full. For an engineer, the glass is already twice as big as it needs to be. An engineer will make do with whatever he can do with the situation he/she is faced with to just make things better.

In this book, we will start off by talking about the personality traits required by an individual and how to work on them to be a successful engineer. Then, we will talk about the challenges and obstacles faced by an engineering student and a working engineer. Since these are based on the personal experiences of the author and his friends, it would be much easier for you to relate to them. From this book, you would also learn things you can gain through taking up engineering degree and being an engineer and how these learning would help you out later on in life.

1. EXCELLENT IN SCIENCE AND MATH-RELATED SUBJECTS

I think I would probably need to start with this aspect on things to know about becoming an engineer. You got to be excellent in science and math-related subjects such as calculus, chemistry and physics. This is because you would be using all this knowledge in real-life engineering applications later on in your working life. This is why an engineer is known as the jack of all trades since he/she is required to be versatile in a lot of subjects. If you have zero interest in these subjects, do not consider taking up engineering course because of peer pressure or any of the elders told you so because you are the one who is going to take up that course and you would not want to waste your four years studying something that is not in your interest at all. It is my responsibility to tell you this because there were some of my batch mates who took up engineering because of parental or peer pressure and they ended up hating it but had to continue because they were left with no choice.

2. PROBLEM-SOLVING SKILLS GOT TO BE TOP-NOTCH

The problem-solving skills of an engineer got to be top-notch. This is because becoming an engineer usually requires you to be in the field and not in the office. The problems that come from the engineering site vary according to the times. For example, if you are working as a mining engineer in a mine, there is a possibility that the blasting cannot be done in the initially designed blast hole because there is a blockage 10 meters into the ground so how would you fight against nature? You can't, that is why you would need to find another optimum blast hole in that area that needs to be blasted as soon as possible with a minimal budget. This is because if the blasting is not done quickly enough, there would not be enough production for the mining company and that blame would fall on you eventually as well.

3. SPONTANEITY IS THE KEY

One thing about becoming an engineer is to be as spontaneous as possible. A lot of engineering students are used to do last-minute work. This attitude is not encouraged to be done in the long run because it would eat up your health but that particular skill to deal with pressure for last-minute work and submitting the assignment right on the deadline is essential when working at sites. Since most of the engineering jobs require you to be on sites, the production plants can go into a mess at any point and a lot of these production plants are old because replacing them are very costly and the companies would not opt for that option due to ensure that they can keep the cost as minimum as possible. If you are working in a company where a lot of equipment used is old, the maintenance issues would be on a rise as well. Therefore, you got to be quick on your feet to solve the maintenance issues because if you don't fix them quick enough, the production is going to be halted and your employers are definitely not going to be happy with that.

4. ATTENTION TO DETAIL

Engineers need to be very attentive and be thoroughly detailed in their observational skills. This is because mayhem would be caused at the workplace if an engineer misses a small minute detail in the problems that arise in the working environment. The small details are usually the one which causes things to go haywire in a lot of places. For example, when I was undergoing blasting internship, I learnt that there was a blasting accident which happened a few years ago in that particular quarry which made the quarry to be closed down for a year. When probed further, I got to know that the quarry engineer who is also the shot firer was not thoroughly detailed with the amount of explosives that are needed for each blast hole. This made the workers' to fill up the blast hole with way too much of explosives than the required amount. This made the blasted rocks to fly further than the intended distance. This resulted in two people being dead and eighteen people injured with a lot of cars and factories nearby the quarry to be wrecked as well. This clearly shows if the engineer paid more attention to detail to the amount of required explosives, this incident could have been prevented.

5. PRESENTATION SKILLS

A lot of engineers are usually perceived to be introverts and to have poor communication skills but that is actually not true. There are a lot of extroverted engineers out there but this could be because they were grilled and roasted in the engineering schools. I did not mean it literally thou, I just meant engineering students are usually pushed into giving out presentations after presentations throughout their entire degree life. This makes those who have the fear of public speaking to overcome their fear as well. In engineering schools, there are a lot of extra-curricular activities which are conducted by a variety of clubs and societies. I would strongly encourage you to join all those activities to develop your interpersonal skills and communication skills. This is because strong communication skills and interpersonal skills are essential in becoming a successful engineer later on since you would be becoming a leader in that organization you are working for. A lot of engineers will be involved in top-level management later on in their career path. Therefore, these skills are best developed in engineering schools rather than waiting to improve on them later on in working life. The working life of an engineer is quite busy which would make it harder for you to find extra time to improve on these skills. Therefore, as mentioned before, do involve yourself in a lot of events and activities in your engineering school. In addition, you would also develop your leadership skills through these participations and get some time off from the stressful studying environment in engineering which brings me to the next point.

6. CAPACITY TO THINK OUTSIDE OF THE BOX

One of the essential things to work on to become an engineer is your capacity to think outside of the box. Engineers are basically classified as creators and problem-solvers. Therefore, this classification requires them to be creative and innovative. I would suggest you to indulge in more events which would make use of your thinking cap. For example, there are a lot of events out there which involve you to create environment-friendly model buildings, bridges using straws and other concepts. You can join in all these events to get a first-hand experience on how engineering life would look like and how you would come up with solutions with a minimal budget and limited time span.

7. CRITICAL THINKING

Being a critical thinker is as important as the capacity to think out of the box for an engineer. Critical thinking would help you divide the thinking process into stages and you would remove the ridiculous ideas right away into the trash. Another thing is basing your decisions based on critical thinking would give assurance to your bosses that you can solve the problems in a proper manner and not a madman attempting to close down their company once and for all. In addition, having this skill would help you to not deviate too much from the standard operating procedure practised in the corporation. Other than that, possession of this skill means you will be reviewing the data and information obtained from past and present before coming up with a decision which would make your decision to be more reliable and trustworthy to be executed.

8. PRACTICALITY

In order to be an engineer, it is important to be creative and innovative but it is very important to be as practical as possible as well. This is because innovation without practicality would not work out in real-life situations. Engineers have to think out of the box and also think about the practical side of it to make sure the ideas could be executed with a minimal budget and the safety standards are not compromised. For example, if you are a civil engineer and working on a project which involves building bridges, the architect and you can execute the innovative and complex design for the bridge but the question is does the design sound logical in real life application and is it practical to be used by people on a daily basis for many years to come? Therefore, it is very important to strike a balance between being practical and being innovative. Innovation is supposed to ease people's burden and make their lives' easier, not putting more burden on them and putting their lives' at stake.

9. BEING A VISIONARY

When we discuss about innovation and practicality, we need to agree on how it corresponds to being a visionary. Visionary is basically someone who can think ahead in the future and anticipate the risks and side effects of the decisions he/she takes right now. Engineers need to have this personality trait because they will usually be working on projects which would be used by others for many years to come. Therefore, they need to anticipate the pros and cons of any decision they take and plan the solutions accordingly to the problems that might arise in the future. Besides, when an individual possesses this personality trait, the individual can be successful in any other industry he pursues later on his/her life. This might explain the fact on why the CEO of many of the major companies in the world are engineers. Some of the most prominent CEO's in the world with an engineering background are Jeff Bezos, CEO of Amazon; Satya Nadella, CEO of Microsoft Corporation; Virginia Rometty, CEO of IBM; Sundar

Pichai, CEO of Google Inc; Dennis Muilenburg, CEO of Boeing and many more. The list just goes on and on.

10. OVERANALYZING

Engineering studies requires you to analyze a lot of the concepts and facts in the syllabus in order to understand them in a detailed manner. That is how engineering syllabus is oriented and arranged. This would make you automatically overanalyze a lot of other aspects in your life and tell others about those aspects to other people in a much more detailed manner compared to how you did it before. For some people, they would be appreciative of you telling them things in a detailed manner but for others, they would find it very annoying to deal with you. Therefore, it is important to strike the right balance in analyzing things and telling it to others in order to get your message across to other people efficiently and without annoying them or making them bored to death.

11. ADAPTABLE TO ANY ENVIRONMENT

Since the engineering degree's syllabus is so versatile with so many components in it, the engineering student is supposed to be versatile as well to undertake all the subjects and graduate through it and versatility requires adaptability. This is because even in high school, you might have some subjects you love more compared to other subjects. This would haunt you in engineering if you do not have an adaptable approach in university. In university, engineering subjects would comprise of subjects you love and subjects you do not prefer studying as well and you need to pass all the subjects in order to graduate. This would help you later in your working environment. Even if you work in a working environment outside of your field, you would still more likely survive and adapt to it because engineering has already taught you how to adapt to any situations and environments and how to be versatile in all situations.

12. PERSEVERANCE AND STRONG WILLPOWER

Engineers and engineering students are constantly up against a wall. In order to be adaptable and versatile to various situations, it requires the individual to have strong willpower and perseverance to pull out through situations which are deemed impossible by others. Engineers are individuals who are capable of thinking of new ways and new perspectives on solving an issue and new ways of thinking always end up getting a lot of resistance. This is why it is important for you to not give up when you have thought about new ways to solve an issue. If they go against it, then just be stubborn and put in the extra hard work required to solve that issue.

Action speaks louder than words, so it's better to show the results first before talking a lot about whether it will work or not. Other than that, perseverance and strong willpower is essential for completing an engineering degree and being an engineer out there because you will see a lot of your batch mates dropping out of engineering after one or two years of taking up the degree because of the difficulty of this course. Whenever you feel like dropping out together with them, always think back on what motivated you to take up engineering in the first place. That would be more than enough to push you back in completing engineering degree and become an ultra-successful engineer.

13. SYSTEMATIC TIME-MANAGEMENT SKILLS

If you are a person who is very active in extra-curricular activities, this habit will definitely interfere in your studies since you need to spend extra time out of your degree to be involved in activities and events. Therefore, it is important that you have systematic time management skills to ensure that you are not left out in your studies too much and also successfully participating in the activities you love doing as well without eating up your physical and mental health. Other than that, possession of this skill would help you later in your busy working life later on and you would not have mental breakdowns when you are constantly faced with insane amount of pressure in your working life.

14. TEAM PLAYER

Engineers are required to be good team players because when you are working as an engineer for a huge corporation, you will be working with a lot of other engineers and also people from other departments depending on your job specifications. In addition, when you are undergoing your engineering degree, there will be a lot of group projects you need to participate in as part of your coursework. If you are an introverted person and do not like socializing that much, I think it is time to tweak that personality a bit if you wish to become an engineer. You cannot run the engineering tasks individually once you are in your working life. Therefore, it is important for you to build your personality as a team player and improve on your capacity to work in teams rather than only working as an individual. Other than that, when you are working in various groups for different projects in university, you get to meet various individuals with different personalities and learn a lot of things about them and yourself as well. This will eventually help you in the real world since your cliques at your workplace would have their unique set of personality traits and you need to work with them to make the projects you are working on to be successful.

15. PROCRASTINATION IS A HABIT TO AVOID

Procrastination is a habit a lot of humans possess although they know it would affect their quality of work and overall health later on. In engineering, the students tend to procrastinate because the amount of tasks handed out sometimes are ridiculous and the deadlines are tight but then, the students tend to get bored and stressed out with the tasks and this leads to procrastination and completion of tasks at last minute. This would affect their quality of work and make them get sick as well. This is a habit that needs to be avoided by the engineering students because the working life of an engineer later on does not give you the same privilege of procrastinating with your problems and tasks assigned to you. If you procrastinate in your working life, you will end up getting fired by your employers for not being efficient enough and that would obviously affect your overall work reputation and career path progress in your life as well. In order to reduce or to avoid this habit, you could write down and break the huge task handed out to you into simpler tasks and set deadlines for the simpler tasks. This is because it is always easier to finish the simpler tasks in chunks rather than looking at the huge task at hand. By the time you finish off the tasks that are segregated into simpler tasks, you would have got a lot nearer to finish off the huge task at hand.

16. PREPARE TO TAKE CALCULATED RISKS

As an engineer, problems would not come to you in advance. The problems in the working environment of an engineer would come all out of sudden and you need to find solutions as quick as possible to make sure the production can start again as soon as possible. The more the amount of time an engineer takes to solve the problems in the working environment, the more the losses that will be acquired by the company. Your boss would not like that, would he? Therefore, an engineer has to be prepared to take calculated risks as soon as possible in a lot of situations. In order to be prepared to take calculated risks at your workplace, you need to prepare yourself years ahead which means in your university life itself. In your university life, when you participate as high committee in events, you will be bound to take a lot of calculated risks before the event takes place due to the deadline of the event. If the risks do not pay off, your whole event could be in jeopardy so that could be the best way to practice taking risks because the events you organize in your university are small-scale risks and can be solved quickly. This would in turn prepare you to take the huge calculated risks in the workplace.

17. LEADERSHIP

Engineers are required to be leaders once they go into the working environment. This is because you would be handling all the laborers who are working at the site and you need to be an effective leader to make sure they carry out their tasks efficiently. This is why it is important that you take up leadership positions in the events you participate in university and high school to sharpen your leadership skills. Leaders are bound to face the brunt of the mistakes that the people working under him/her make. Therefore, this would help you to learn from your mistakes and those who are working with you as well. Other than that, working with other people who could be your friends also helps in having a better understanding about those individuals as well. An individual's real character shines through when he is under intense pressure.

18. KEEN ON TECHNOLOGY

As an engineering student, if you are not passionate and keen on learning about technology, you are not an ideal fit to be an engineer. Technology is the major part of what is engineering in this today's world. Therefore, it is essential that an engineer needs to be very interested and passionate about topics related to technology. If you do not update yourself on new information on technology, the company you are working for might not even know about new machines and new ways of carrying out their tasks. These new machines and new tasks would probably carry out the tasks they are currently executing at a lower cost and lesser time.

19. CONTINUOUS LEARNING

The world is constantly changing every day. New technologies are coming in every single day and new problems are arising on a constant basis. It's the job of engineers to make sure they are updated with all these new information to find out ways on how to solve the current issues they are facing in a much easier manner and at a much lower cost. This is why continuous learning is very important. Your thirst for education must not stop at high school and university but must be continuous till your death. As an engineer, you need to just read a lot on new updates on various issues to have better communication with your clients and bosses and to have a deeper understanding on what you can do to improve the world around you. Engineers are not professional slaves to be only working for a company; they are the individuals capable of changing the landscape of the world to let others to live a better life with higher quality.

20. CURIOSITY

Curiosity killed the cat but that is not the case with engineers. The more inquisitive you are, the more answers you will end up finding. As an engineer, you need to be curious on a lot of things. For example, an engineering student or an engineer need to ask a lot of questions about the machine he just saw and research them on the internet on how it works and to have a better understanding of the machine. By having a better understanding, you can think of ways to utilize the machine in a more effective and efficient manner. This is how you can allow yourself to have better understanding on things you work with in your working environment and think of better ways on improve the working environment. This does not limit to your working environment only. When your habit of curiosity goes into thinking about solving the problems of people around you, you help create solutions for them and improve their lives' as well.

21. HONESTY IS THE BEST POLICY

Honesty is a trait that engineers should hold close to their hearts. Since you would be usually working in groups in the life of an engineer or an engineering student, it is very important to maintain honesty between your group members or cliques. You should not hide your mistakes when the situations do not favor you. If you hide them and try to solve them on your own, you might just end up making the situations even worse not only for you but for everyone working together with you. Trustworthiness is what keeps the team united and result in productive outcomes. A lot of projects have failed miserably due to dishonesty since dishonesty in engineering usually means problems and issues that arise are not addressed in a proper manner and ended up not being solved at all.

22. BE HUMBLE

Many engineering students make mistake by being arrogant of their past achievements in high school. If you are selected for an engineering school, the probability of you being excellent in mathematics and science are very high. Let me tell you something, engineering is not a walk in the park so drop your arrogant attitude and stop basking in your past glory way too much. Humility is the way to survive in the engineering world. The attitude of being humble and acknowledging that you need to keep learning throughout your engineering degree would help you to maintain your GPA. Trust me, the GPA of the first two semesters are very important in maintaining a high overall CGPA and I regretted not making the best out of it. I always revised for my exams at last minute in high school and end up getting excellent results and thought I could nail the exams in engineering degree in the same way. Oh boy, how wrong was I! I failed in four subjects in the first two semesters and my CGPA at the end of the first year was below 2.00. I had to put in extra hard work and have sleepless nights for the next six semesters in order to increase my overall GPA to go above 3.00 in order to get a Upper Second Class Honours at the end of the engineering degree. What a rollercoaster ride it was. Therefore, do not end up doing it like me and focus on your subjects on your first year degree. The subjects in your first degree would usually be focused on the core basics of engineering. Therefore, it is important that you pay attention and dedicate your time to learn the subjects in a more comprehensive way since that knowledge would be important in your working life later on as well.

23. CONSISTENCY

As I have mentioned before, engineering requires a lot of determination and hard work but that hard work should not be put in at the last minute. You need to be consistent throughout the whole semester to nail the subjects. For example, I failed my Engineering Mechanics subject in my first semester due to my lack of consistency in revising for the subject. In the first attempt of retaking that paper, I made schedule to study the subjects and set deadlines for each subtopic to make sure I can finish revising the subject on time before the paper. I managed to finish my revision and ended up getting excellent result in that subject.

Consistency is the key if you would like to finish your engineering degree in four years. Other than that, consistency would help you out to be more flexible just in case you need to go out for hangouts or for any other matters since you have already finished the assignment or revision ahead of the planned time.

24. INTELLIGENCE ALWAYS HAVE TO COME OUT ON TOP

There are a lot of instances where our mind can be muddled up with emotions from the problems of our personal lives'. You should never let your emotions to win over your intelligence as an engineer. For example, imagine that you are a civil engineer and you are handed over the responsibility to build high-rise apartments. What do you think would happen if you let your emotions to interfere in your line of work? It will definitely interrupt your focus on the project and look at how many people's lives are at stake if you are not focused on your work. This is the reason why it is so important to not let yourself lose yourself in your emotions because you can literally put so many people in danger.

25. SOCIALIZING

Socializing is very important for an engineering student or in the working life of an engineer. If you are introverted, do have two or three close friends who you can talk to on a frequent basis. Engineering can make anyone go crazy with its tight and busy schedule so it is important that you have people you can rant to about your problems. If you keep too many issues and problems inside you without sharing them with anyone else, that would eat up your mental health. When your mental health is in complete mayhem, it will definitely affect your physical health as well. You do not want to end up being a disease-prone person at a young age itself, right? Therefore, do socialize and get your problems right from your mind.

26. BUILDING CONNECTIONS

When it comes to socializing, another important thing to remember is the part where you build more connections. For example, if you are participating in a lot of events, you will get the opportunity to meet more people from other engineering disciplines and even other successful people in their career pathway. For example, I got the opportunity to communicate with a few successful people through the Ted Talk I attended and got inspired to do what I love no matter what obstacles lie in my pathway. The aspect of building connections will help you once you are in your working life to discover new opportunities. Who knows, perhaps one of the friends you made from another engineering discipline in an event you participated in could be a CEO of a company in 5-10 years' time and could be calling you up on becoming his shareholder just because you are his friend? You never know the possibilities but there is a small probability that things like these might happen if you have a lot of connections. You do not know when you would require a certain help from a certain individual. Therefore, do build a lot of connections throughout your university life and even in working life later on to not miss out on any possible opportunities out there.

27. ENGINEERING DEGREE IS INDEED HARD

Well, by now, you would have seen a lot of engineering memes circulating in social media on how messed up engineering studies are. Unfortunately, it is a true fact and engineering studies are indeed hard because the subjects in the engineering degree syllabus comprises of advanced levels of the most of the subjects you undertook in high school such as Chemistry, Physics, Mathematics and many more. There are even subjects discussing about biology in Engineering. It is hard but do not worry, that is why I am here. The deadlines of the assignments in the engineering degree are very short as well which would force you to work around the clock. The way to work around all these is to list down things needed to be done and their deadlines so that you can assign and finish the tasks based on priority basis. In addition, you can conduct revision session with a friend of yours because two heads are better than one. You can have three or more heads to revise together as well but do not indulge in group study sessions with too many people because too many cooks will absolutely spoil the broth.

28. PACE IS WAY TOO FAST

One thing about engineering subjects is that each subject has a lot of chapters in it. For example, one engineering subject might comprise of 20 chapters in it and you might need to take five to six subjects like that per semester. In those 20 chapters, there are a lot of subchapters in it as well and mind you, one semester is about four to six months long and all these chapters have to be completed by then. Therefore, the teaching pace in engineering degree would be very fast so if you didn't pay attention for even one or two weeks, you will be left with too many things to catch up on which would make you feel like a miserable person. This makes it important to make sure you are frequently attending your classes to take note on where your class is at regarding the syllabus and you could prepare for the next chapters accordingly in advance.

29. REDUCTION IN GUARANTEE FOR JOBS

The situations and job opportunities have changed now. Engineering jobs are becoming harder to get because the amount of engineering graduates per year is on a rapid increase but the amount of engineering jobs are not sufficient to provide job placements for every single engineer out there. For example, there was a senior of mine who waited for eight months to get a job in her engineering discipline although she graduated with a first class honors engineering degree. That's how difficult things are getting right now. Therefore, if you think engineering degree would easily land you a job right after you graduate, you can drop that thought. Take up engineering only if you are absolutely sure that you are very passionate about it and not because you want to have job security. There is absolutely reduction in guarantee for engineering jobs under the current circumstances.

30. SETTING LOW EXPECTATIONS ON SALARY

In engineering job, job guarantee is one thing and salary is another thing. Many people have a perception that engineering is a high paying job but that is not the actual case over here. As a fresh graduate, you do not earn much and sometimes your pay would be lesser than those fresh graduates from other fields such as Human Resources, Finance or Management. That is how the corporates have set the salary for the engineers although I believe our workload is much heavier due to the nature of our job which requires engineers to work on sites and come back to the office and do the paperwork as well. As I have mentioned before, passion is the one thing that would drive you in this engineering field more than anything. The amount of pay a fresh engineering graduate might be low, but as you gain more experience, you would be getting promotions much faster which would result in higher pay. This means your rate of increase in pay rate as an engineer is going to be much higher than a lot of other people working in other fields. Patience is a virtue and therefore, you need to be as patient as possible and wait till you gain sufficient experience to get better job opportunities with better pay.

31. GPA DOES NOT MATTER AS MUCH AS YOU THINK

Nope, I am not contradicting myself here. I did mention that maintaining a good GPA is important but it is not the only important aspect in your life as an engineering student. GPA obviously becomes a topic of worry for many engineering students because that is one of the most important criteria in securing your first job. Your resume will be highlighting that criteria in bold to the employers so a lot of engineering students would be high on caffeine intake to make sure they cover the engineering syllabus as much as possible. Let me tell you something. GPA does matter but it does not matter to the employers as much as you think. It is used as a criterion to filter out the candidates for interview phase. You need to have excellent interpersonal and communication skills to ace the interviews for the job you applied for. Therefore, do balance out your time in acquiring other skills in your life as an engineering student. For example, I participated in a lot of events and became

41

Head of Sponsorship for an event and Head of Project Management for another event. I gained varied experience during my time working at different departments since the job scope for both the departments are varied. This made me to be more versatile and improved my flexibility on managing other tasks in my life. This particular experience and skills would help out later in your working life when your job scope requires you to be excellent in other aspects other than just engineering. You can enroll in other courses during your life as an engineering student such as taking up foreign language class. You have no idea when the knowledge of knowing an extra language would help out in your life. On top of that, you can get enrolled in other computer-related skills which are not related to your engineering discipline. For example, CATIA software is not part of mining engineering syllabus but some of my friends did take it to gain extra knowledge on other types of software. Programming is a subject that you can basically learn on your own if you are interested in it since there are so many free courses related to programming which are offered in internet. One such website that offers free courses related to programming is Coursera. The extra knowledge you gain outside of your engineering syllabus will never be wasted and at one point in your life, it will definitely come in handy.

32. CONSTANTLY BEING ASKED TO FIX THINGS

When undergoing an engineering course, you will be usually doing specialization courses such as mechanical, electrical, chemical, robotics, mining and etc. I did specialization in mining engineering so I would not really know what the other courses are about in a detailed manner. Generally, I know what their course is about because you will learn everything about the basics of engineering in your first year of engineering degree. You and I know that but the people who did not take up engineering course would not know that. You might be specialized in mining engineering but your uncle could come and ask for help to fix the short circuit in his house. That is how things are once you become an engineer. Well, it is part and parcel of engineering life to be constantly asked to fix things although a lot of the things they ask to fix do not come under your specialization.

33. BEING ASKED RIDICULOUS QUESTIONS

Other than being bugged with constantly asked to fix things, you are going to be asked ridiculous questions as well. There was a friend of mine who did chemical engineering and one of her mother's friends asked her if she knows how to make bombs, literal facepalm. I was also asked the same question before since I did my internship for a blasting company at a quarry site so my job is basically blasting rocks to get optimum fragmentation so that further processing can be done on them. Engineering degree was supposed to create engineers, not terrorists and some people just don't understand this, I guess.

34. FEELING STUPID MOST OF THE TIME

You might be labelled as a complete genius in high school due to your way of tackling various subjects of different fields with ease but then, you might feel like a complete idiot once you enter university. The quick pace of teaching as mentioned before and the complexity of each of the subjects might feel like you have grown stupid and you might feel like you are the only stupid person out there. Trust me, you are not. About 90-95% of the engineering students feel like that and obviously, the other 5-10% are complete geniuses so let us leave them out of this discussion. It is okay to feel stupid but do ask questions frequently to people who know the topics better. There are no stupid questions in engineering and most of the lecturers in the university are always willing to help out their students with their studies. You just need to gather your courage and ask the questions that baffle you up.

35. FINAL YEAR PROJECT IS A COMPLETE MAYHEM

One thing you need to understand about engineering degree is it is full of obstacles but nothing comes close to the mayhem caused by your final year project. Final year project is generally a project undertaken by a student which usually specializes in a particular topic in his/her engineering discipline. There will be problems throughout the journey of your final year project no matter how well-prepared you are for the final year project. For example, I was very well-prepared for my final year project but my sample from the quarry site arrived a month late compared to the scheduled date which means I have to squeeze in that one month of scheduled work in the remaining time left which made me to have intense pressure and have sleepless nights due to the stress. If you are doing experimental-based projects, you are in for more troubles ahead. Experimental-based FYP means you are dealing with something outside of the classroom and you are putting the things you learnt throughout your degree to use in real-life application. Some of the universities have state-of-the-art facilities which would ease the burden of students' when it comes to equipment maintenance and breakdown issues. The problem is a lot of universities out there use old equipment and they would not change due to the high cost of the new equipment. Therefore, you will face setbacks and would not get accurate results and that would not be your fault in the first place but you will end up getting scolded by your supervisor for having messed up results. It is usually the same thing when you enter working life later on. If the production is not delivered right on time due to machine breakdown at the site, you will be held responsible for it as well. This is where the personality trait of being able to think out of the box comes in. For example, when I was conducting my final year project, the valve and the pipe that supplies the carbon dioxide from the tank to the mixture was faulty. Therefore, the experiment went completely haywire and I ended up not getting accurate results. I decided to come up with a creative solution for that. I let the carbon dioxide gas through the valve and pipe and sprayed bubble water throughout the pipe to see where the faulty parts are and taped it shut. Then, I started getting accurate results. The worst problems usually can be solved with a simple solution.

36. UNSTEADY RELATIONSHIPS

Being an engineer means you would be way too busy with your work because engineers in current circumstances are required to carry out tasks outside of their job scope at times because the companies are cutting down on the appointment of the new workforce. Therefore, you got to be ready for the unsteady phase in your relationship with your partner.

Even as an engineering student, you are bound to have a rocky phase in the relationship once you enter engineering because of the culture shock you would suffer after taking up this course. Many of my friends have undergone breakups after taking up engineering because they could not cope up in balancing their relationship and engineering life. The base of every relationship is understanding, trust and loyalty. If you would wish to be in a relationship while undergoing engineering, make sure you find a very understanding partner and someone who would understand why you are not spending enough time with him/her.

37. SLEEP CYCLE IS GOING TO BE HAYWIRE

Since you would be working on tight deadlines on a constant basis, your sleep cycle will take a beating. When you go to an engineering hostel, you would be seeing the lights of a lot of rooms out there would be switched on till 4-5 a.m. Then, they would be in classes at 8 or 9 a.m. This is the case with a lot of engineering students out there. This is because a lot of them do their work at the last minute even on those assignments with long deadlines and end up not having enough time to finish up their assignments. Therefore, they have to sacrifice their sleep. Well, since you already know you will have limited sleeping hours once you enter engineering, it is better you start sleeping early and waking up early. One of my friends who was also in engineering sleeps at 11 p.m. and wake up at 5 a.m almost every day and had a very productive life. He usually passes his exams with flying colours and had a great social life as well. He excelled in volleyball and badminton as well throughout his time in engineering degree and was involved in competitions as well. This is an example of the early bird gets the worm and how productive you can get if you have the same sleep cycle every day and if you did it earlier than others.

38. CAPACITY TO VENTURE OUT OF ENGINEERING

Since the life of an engineering student makes you to be involved in a lot of stuff, this trains you to be versatile in a lot of aspects as well. This means you are going to acquire a lot of skills which can be used for other jobs as well. This could be the reason why you see a lot of engineers who graduated with a degree in engineering are able to succeed outside of their field although it is completely out of their job scope. For example, there are movie stars who did engineering and currently being A-list actors in Hollywood. One of them includes Rowan Atkinson who is known worldwide as one of the most remarkable characters ever created for TV sitcoms which is Mr. Bean and also for his role in Johnny English series. He graduated with an electrical engineering degree from Newcastle University. Another movie star who has a background in engineering is Ashton Kutcher who graduated with a degree in Biochemical Engineering from the University of Iowa.

39. YOUTUBE TUTORIAL VIDEOS ARE COMPLETE LIFE-SAVERS

Since the pace of engineering syllabus goes way too fast, it is normal to not be able to catch up with the studies especially if you are a slow learner. Do not worry because there are so many Youtube channels out there which offers free tutorial lessons dedicated to engineering students. For example, PatrickJMT and Khan Academy are two of the prominent Youtube channels visited by me and my friends during my time as an engineering student. If you cannot grasp the concept that is being taught in one channel, you will have ample of other options available for you to go and nail the subject. Majority of engineering students prefer watching these videos because it is free of charge, easily accessible and you can access it at any time you want.

40. LAB TIMES ARE VERY IMPORTANT

Lab work and lab-related subjects are the ones where you learn about the practical side of engineering and allow you to go get hands-on experience. Therefore, do a lot of reading through the internet regarding each topic you are going to carry out the lab work before going into the lab. This would help you to carry out the lab work in a better manner and you would understand the purpose of conducting each and every lab work if you research it thoroughly before entering the lab class. Other than that, putting in extra effort in conducting this lab work will help out when you are conducting your final year project later on. Lab works are usually group sessions so use that opportunity to learn since the final year project is conducted by an individual instead of in group sessions.

41. EATING DONE RIGHT

I know you will get busy with your studies when you are an engineering student or when you are working as an engineer. This would definitely interfere in your eating times but this should not be the reason for you to not eat well. The moment you don't eat well, you will lose focus and your productivity rate will decrease. So, as an engineer or engineering students who are required to change people's lives, you need to eat right and have good nutrition. Make sure you don't skip meals and eat proper amounts of meals at each mealtime. We do not want you to lose your productivity rate and focus and make bad decisions which will, in turn, increase the burden on the lives' of others so make sure you eat right. Consult a dietician if you are confused about proper eating habits and they will advise you on how to do it. Lastly, avoid taking way too much energy drinks since they are very high in sugar and would only lead to health problems such as diabetes if you have an excessive intake of it.

42. DO NOT BE AFRAID TO TALK ABOUT YOUR MENTAL HEALTH

Your mental health will take a lot of beating throughout your life of being an engineering student and engineer due to the demanding nature of the industry and the versatility required of an engineer. This can be very taxing for some individuals and would be worse off for people who are facing other personal problems in their lives.' Therefore, whenever you are facing such issues, you should not be afraid to talk about your mental health issues. The best option is to talk to a psychiatrist or a counsellor when you are facing mental health issues so that they can diagnose you and provide you with further treatment on the issues you are facing. Do have some close trusted people in your circle where you can talk about your mental health issues without being judged and do not care about society's perception of this particular issue. At the end of the day, they are not going to live your lives' in your shoes, you are the only one capable of living your life.

43. LOOK FOR ENGINEERING INTERNSHIPS DURING YOUR SEMESTER BREAKS

I know semester breaks are meant for rest after that tiring semester you went through and the compulsory internship period will only be once in your engineering degree as an engineering student but hear me out. Internships are where you can get to know how your future working environment is going to look like. In most universities, the compulsory internship session would be between a period of three to six months within that four years of engineering degree which is actually very less. Therefore, you can go out and seek internship opportunities in your first year itself in major companies. Most of the companies would be very happy to take in interns because the salary paid to interns are minuscule compared to hiring a graduate. The more time you spend in undergoing productive internships, the more adaptable and knowledgeable you will be once you finish your engineering degree. You will be ahead of the others due to the amount of experience you have under your belt even before progressing to the working life. There is a reason why I said productive internships. There are some internships where there is no planned schedule on what to do and when to do anything. These internships usually make the engineering student to just make coffees, make photocopies of the documents in the office and all the other office work that is completely not related to engineering. Therefore, make sure you take up an internship opportunity where the company has a planned working schedule for an intern. In that way, you can gain proper knowledge of the industry you are going to indulge yourself in a few years' time.

44. ELECTIVES ARE NOT THAT BAD

Engineering degree offers electives and made it compulsory for students to take up a few electives throughout their journey. A lot of students dread elective subjects because they feel it is an extra burden besides all the core subjects they are required to take up. Electives are actually not that bad because it gives you an opportunity to unwind and study other aspects besides engineering. For example, I took modern dancing as an elective and managed to learn the basics of contemporary dancing. Obviously, this has nothing to do with engineering but knowledge is wealth. Who knows, perhaps I am required to do contemporary dance in my best friend's wedding in the future and I could just do it because I already have the basics in it. You never know, the possibilities are endless. Other than that, taking up elective subjects helps you become a well-rounded person and helps you in surviving in other working environments if you did not choose engineering as a profession in the near future.

45. TAKE ADVICE FROM YOUR ENGINEERING SENIORS

I know how it feels like to be lost in the world of engineering. I have been there before and that is the time you require guidance. When you are in doubt of the ways to tackle the problems you face, ask for guidance and advice from your engineering seniors. These engineering seniors have gone through similar sorts of issues that you have faced before. You should not be shy to ask for help from these seniors because most of them would be willing to help out their juniors. I do agree that some of them are snobbish and arrogant but many of them are cool individuals who would be glad to help you out.

46. YOU WILL BE SEEING A LOT OF SNAKES

The term snakes here that I meant here is two-faced humans who would backstab or betray you. Therefore, it is important to remember that you

59

should not place extreme amounts of trust on other people when you enter the world of engineering. Since the environment is harsh and hyper-competitive, there would be a lot of problems that are faced by individuals especially in group projects and many things can go wrong. People would start the blaming game when things go wrong and maybe at one point, you can be blamed on a mistake that you did not commit by your own 'trusted' best friend. That is how harsh the environment of engineering is. Therefore, be cautious when making friends out there as an engineer or an engineering student and avoid having too much trust in people.

47. ATTENDANCE RECORDS ARE IMPORTANT

Even though you are someone who can revise the particular subject on your own and do not require going for classes to understand the topics, please do attend the classes on a consistent basis. Attendance records are very important in engineering degrees and you can be barred from taking the finals if you do not have enough attendance recorded for the classes. In addition, consistent attendance of the classes will help you out in understanding the pace and nature of the subject taught. These professors are usually PhD and Masters holder and you can learn so much stuff on engineering and life from them, therefore do not miss classes. Other than that, having a high attendance record in your university degree will help you out later in your working life. It would help you out in having a consistent attendance at the workplace without you dreading about it since you are already used to that lifestyle for a long time. Imagine having bad attendance records at the workplace. You can be fired from your job for not being disciplined enough or your productivity in finishing a project can decrease a lot because you have not been spending a lot of time on your project.

48. BUILD A STRONG RAPPORT WITH YOUR PROFESSORS AS AN ENGINEERING STUDENT

The habit of building a strong rapport with your professors during your engineering course is very important. This is because they can help you and guide you better in your studies when you have a strong rapport with them. Professors usually have a lot of connections in the engineering industry you are going to be involved in the future. This will help you when you are looking for job opportunities after completing your degree. If you have a great relationship with them, they will easily become your person to contact under recommendations in your resume and they will definitely give positive feedback about you to the company who is thinking of hiring you if you build a strong rapport with your professors.

49. TAKE FAILURES AS A STEPPING STONE TO SUCCESS

You will screw up at some point in your engineering life no matter how successful you were at everything before and it is completely okay. Everyone who enters the engineering degree will screw up at some point and they all have risen back to achieve the pinnacle of success. It is okay to fail your paper in the first time you take the paper although you have put in full hard work to pass the paper. You can always retake it. I would strongly suggest you to not fail a particular paper on purpose based on what I said just now because retaking the paper would mean you need to put in extra time and effort besides the subjects you take up in the upcoming semesters. Please do put in a lot of hard work for each subject you are taking up in your degree and if you fail them after that, take those failures as a stepping stone to success. As I have mentioned, final year project will also result in a lot of failures and disappointments in the initial phase and sometimes, even in the final phase but engineering is all about improvisation. Therefore, take your time to step back and think deeply to create solutions for the failures in your engineering life instead of overthinking on your failures. In the working life of an engineer, you would not have the privilege of overthinking too much on your failures. As an engineer, you need to rectify those mistakes as soon as possible to prevent the working environment from going into a deeper mess.

50. LEARN CODING AS MUCH AS POSSIBLE

You need to learn coding as much as possible in your engineering degree. In this modern era, coding is what makes up about most of the stuff that is possessed by our electronic devices. I think all the engineering schools require you to learn the basics of coding and usually, the programming language used as part of engineering languages is the C++ programming language. There are some websites which offer free coding classes which include Coursera, Codeacademy and Udemy. Coding will help you in developing your creative and innovative ideas that pop out of your minds into new applications which will help out others in a lot of aspects.

BONUS TIP:

Avoid Indulging In Habits That Will Jeopardize Your Health

The habits I meant was smoking, vaping and binge drinking on a consistent basis. Smoking and vaping do relieve stress, but in the long run, it will affect your health and make you regret your decision to take it up in the first place. Smoking and vaping are two habits that cannot be left easily and many of my friends who tried to abstain from the habit completely failed in their attempts. Drinking alcohol is fine but binge drinking due to the stress of engineering life is bad and you should not be doing it on a consistent basis especially not right before the exams. You do not want to sit through an exam you prepared for with hangover. The worst possible outcomes coming from it are endless so avoid doing that.

OTHER HELPFUL RESOURCES

1) The mistakes to be avoided throughout engineering college life.
https://www.quora.com/What-are-the-mistakes-that-I-should-not-do-in-my-4-year-engineering-college-life

2) The three essential properties of the engineering mindset
https://fs.blog/2015/06/the-engineering-mind-set/

3) An engineer's thoughts on work/life balance
https://blogs.cisco.com/perspectives/an-engineers-thoughts-on-worklife-balance

4) How to study engineering and still have a life
https://www.engineering.com/Education/EducationArticles/ArticleID/10898/How-to-Study-Engineering-and-Still-Have-A-Life.aspx

5) The Engineering School Survival Guide: 4 Frameworks To Dominate Your Degree
https://collegeinfogeek.com/engineering-school-survival-guide/

READ OTHER

50 THINGS TO KNOW

BOOKS

50 Things to Know

Stay up to date with new releases on Amazon:
https://amzn.to/2VPNGr7

Mailing List: Join the 50 Things to Know Mailing List to Learn About New Releases

50 Things to Know

Please leave your honest review of this book on Amazon and Goodreads. We appreciate your positive and constructive feedback. Thank you.